ASTON MARTIN

The ultimate guide to understanding Aston Martin's origin and history, net worth and car models

By

Dawn Loveys

Table of contents

Chapter 1:Aston Martin company history

British luxury sports cars and grand tourers are manufactured by Aston Martin Lagonda Global Holdings PLC. Its ancestor was established in 1913 by Lionel Martin and Robert Bamford.

The two had united as Bamford and Martin the earlier year to sell vehicles made by Vocalist from premises in Immature Road, London where they additionally overhauled GWK and Calthorpe vehicles. Martin decided to build their own cars after racing specials at Aston Hill, near Aston Clinton. By attaching a four-cylinder Coventry-Simplex engine to the chassis of a 1908 Isotta Fraschini, Martin built the first automobile to be known as an Aston Martin.

They gained premises at Henniker Mews in Kensington and delivered their most memorable vehicle in Walk 1915. When the First World War broke out, Martin joined the Admiralty and Bamford joined the Army Service Corps, so production couldn't start.

They built a new car and relocated to Abingdon Road, Kensington, after the war. Bamford & Martin was revived with funding from Count Louis Zborowski after Bamford left in 1920. In 1922, Bamford and Martin created vehicles to contend in the French Stupendous Prix, which proceeded to set world speed and perseverance records at Brooklands. For racing and breaking records, three works Team Cars with 16-valve twin cam engines were constructed: skeleton number 1914, later created as the Green Pea; frame number 1915, the Extremely sharp steel record vehicle; as well as chassis number 1916, which was later renamed the Halford Special.

In two configurations, approximately 55 automobiles were produced for sale; both short and long chassis Bamford and Martin failed in 1924 and was purchased by Dorothea, Woman Charnwood, who put her child John Benson on the board.

Bamford and Martin got into monetary trouble again in 1925 and Martin had to sell the organization (Bamford had previously left it in 1920).

Later that year, the company was taken over by Bill Renwick, Augustus (Bert) Bertelli, and investors that included Lady Charnwood. They rebranded it as Aston Martin Motors and relocated it to the Feltham facility of the former Whitehead Aircraft Limited Hanworth plant. Renwick and Bertelli had been in organization a few years and had fostered an above cam four-chamber motor utilizing Renwick's licensed burning chamber plan, which they had tried in an Enfield-Allday frame. It was the only "Renwick and Bertelli" automobile produced, and it is still in existence under the name "Buzzbox."

The pair had wanted to offer their motor to engine makers, however having heard that Aston Martin was presently not underway acknowledged they could benefit from its standing to kick off the creation of a totally new vehicle.

Bertelli was the technical director and designer of all new Aston Martins from 1926 to 1937, which became known as "Bertelli cars." They incorporated the 1½-liter "T-type", "Worldwide", "Le Monitors", "MKII" and its hustling subordinate, the "Ulster", and the 2-liter 15/98 and its dashing subsidiary, the "Speed Model". Most were open two-seater sports vehicles bodied by Bert Bertelli's sibling Enrico (Harry), with few long-undercarriage four-seater sightseers, dropheads and cantinas likewise delivered.

One of the few owners, manufacturers, and drivers, Bertelli was a competent driver who loved to race his cars. The "LM" group vehicles were extremely fruitful in public and worldwide engine dashing including at Le Monitors.

In 1932, financial issues resurfaced. Before being given to Sir Arthur Sutherland, Lance Prideaux Brune saved Aston Martin for a year. Aston Martin made the decision in 1936 to focus on road cars, producing just 700 vehicles before the Second World War halted production. Creation moved to airplane parts during the conflict.

1947–1972: David Brown

In 1947, old-laid out (1860) exclusive Huddersfield stuff and machine devices producer David Earthy colored Restricted purchased Aston Martin, putting it taken care of its Work vehicle Gathering. The most recent hero of Aston Martin was David Brown. He additionally obtained Lagonda,without its industrial facility, for its 2.6-liter W. O. Bentley-planned motor. Lagonda shared engines, resources, and workshops when it moved operations to Newport Pagnell. Aston Martin started to assemble the work of art "DB" series of vehicles.

They announced that their Le Mans prototype, which would be called the DB2, would be made in April 1950. The DB2 was followed by the DB2/4 in 1953, the DB2/4 MkII in 1955, the DB Mark III in 1957, and the Italian-styled 3.7-liter DB4 in 1958.

Even though these models helped Aston Martin build a strong racing legacy, the DB4 stood out and inspired the DB5 in 1963. Aston remained consistent with its excellent visiting style with the DB6 (1965-70), and DBS (1967-1972).

The six-chamber motors of these vehicles from 1954 up to 1965 were planned by Tadek Marek.

1972–1975: William Willson

Aston Martin was frequently monetarily grieved. David Brown sold Company Developments, a Birmingham-based investment bank consortium led by accountant William Willson, for £101 in 1972 after paying off all of the company's debts, which were estimated to be at least £5 million. More detail on this period might be perused at Willson's life story. Aston Martin was again forced into receivership at the end of 1974 due to the global recession, a lack of working capital, and the difficulties of developing an engine to meet California's exhaust emission requirements. This put an end to the company's sales in the United States. The organization had utilized 460 specialists while the assembling plant shut.

1975–1981: Sprague and Curtis In April 1975, the receiver sold the company to North American businessman Peter Sprague of National Semiconductor, Toronto hotelier George Minden, and London businessman Jeremy Turner, who insisted to reporters that Aston Martin remained under British control. Later, Sprague said that he had fallen in love with the factory, not the cars, because of the hard work, intelligence, and dedication of the workers. He and Minden had brought in British office property developer Alan Curtis, an investor, and retired Sheffield steel magnate George Flather.

A half year after the fact, in September 1975, the production line - shut down the past December - re-opened under its new proprietor as Aston Martin Lagonda Restricted with 100 workers, and wanted to lift staff to 250 toward the finish of 1975. In January 1976, AML uncovered that it currently held orders for 150 vehicles for the US, 100 for different business sectors and one more 80 from a Japanese bringing in organization. Fred Hartley, who previously served as managing director and sales director for 13 years, announced his resignation at the Geneva Motor Show due to "differences in marketing policy."

The new proprietors drove Aston Martin into modernizing its line, presenting the V8 Vantage in 1977, the convertible Volante in 1978, and the oddball Bulldog styled by William Towns in 1980. The futuristic new Lagonda saloon, based on the V8 model, was also styled by Towns.

Curtis, who had a 42% stake in Aston Martin,also achieved a shift in course from the standard clients who were Aston Martin fans, to effective youthful wedded money managers. There was a rumor that AML was about to acquire Lamborghini, an Italian automaker, and prices had increased by 25%. At the conclusion of the 1970s, there was a lot of discussion about incorporating MG into the Aston Martin consortium. 85 Moderate MPs shaped themselves into a tension gathering to get English Leyland to deliver their hold and hand it over. CH Industrials plc (vehicle parts) purchased a 10% offer in AML. However, in July 1980, blaming a recession, AML laid off 450 employees by reducing their workforce by more than 20%.

1981–1987: Victor Gauntlett Alan Curtis and Peter Sprague announced in January 1981 that they had never intended to maintain a long-term financial stake in Aston Martin Lagonda and that it would be sold to Victor Gauntlett of Pace Petroleum because there had been no satisfactory revival partners. According to Sprague and Curtis, AML's finances had improved during their ownership, making it possible to make an offer for MG.

In 1980, Gauntlett paid £500,000 to Pace Petroleum to acquire a 12.5% stake in Aston Martin, with Tim Hearley of CH Industrials acquiring a similar share. Speed and CHI took over as joint 50/50 proprietors toward the start of 1981, with Gauntlett as leader administrator. After some development and publicity, Gauntlett was able to sell the Lagonda in Oman, Kuwait, and Qatar after it became the world's fastest four-seat production car. The Prince of Wales issued Aston Martin with a Royal Warrant of Appointment in 1982.

They established an engineering service subsidiary to develop automotive products for other businesses because they were aware that developing new Aston Martin products would take some time. It was chosen to utilize a trademark of Salmons and Child, their in-house coachbuilder, Tickford, which Aston Martin had purchased in 1955. The name Tickford had long been associated with high-priced, high-quality automobiles and carriages with folding roofs. Pace continued to sponsor racing events and now sponsored all Aston Martin Owners Club events, taking a Tickford-engined Nimrod Group C car owned by AMOC President Viscount Downe, which finished third in the Manufacturers Championship in both 1982 and 1983. Other new products included a Tickford Austin Metro, a Tickford Ford Capri, and even Tickford train interiors, particularly on the Jaguar XJS. It likewise completed seventh in the 1982 24 Hours of Le Monitors race. Notwithstanding, deals of creation vehicles were presently at an unequaled low of 30 vehicles delivered in 1982.

As exchanging became more tight in the oil market, and Aston Martin was calling for additional investment and cash, Gauntlett consented to offer Feeds/Speed to the Kuwait Venture Office in September 1983. While Gauntlett remained chairman of Aston Martin, 55% of the stake was owned by ALL, and Tickford was a 50/50 venture between ALL and CHI. As Aston Martin required greater investment, he also agreed to sell his share holding to American importer and Greek shipping tycoon Peter Livanos, who invested through his joint venture with Nick and John Papanicolaou, ALL Inc. When ALL exercised their options to acquire a larger share of AML, the tense relationship came to an end; The remaining shares of CHI were exchanged for CHI's complete ownership of Tickford, retaining the development of Aston Martin projects that were already underway. In 1984, Papanicolaou's Titan shipping business was in trouble, so George Papanicolaou,

Livanos's father, bought out the Papanicolaou family's shares in ALL. Gauntlett became a shareholder once more, this time with a 25% stake in AML. The arrangement esteemed Aston Martin/AML at £2 million, the year it assembled its 10,000th vehicle.

Despite the fact that subsequently Aston Martin needed to make 60 individuals from the labor force excess, Gauntlett purchased a stake in Italian styling house Zagato, and restored its joint effort with Aston Martin.[52] In 1986, Gauntlett arranged the arrival of the imaginary English spy James Cling to Aston Martin. Cubby Broccoli had decided to rework the person utilizing entertainer Timothy Dalton, trying to re-root the Bond-brand back to a more Sean Connery-like feel. Gauntlett sold Broccoli a Volante to use at his home in America and provided his personal pre-production Vantage for use in The Living Daylights. However, Gauntlett declined the film's offer to play a KGB colonel: I would have wanted to have gotten it done yet truly couldn't manage the cost of the time."

1987–2007: Ford Motor Company bought a 75% stake in Aston Martin in 1987 and the rest later because the company needed money to last. Victor Gauntlett and Prince Michael of Kent watched the revival in May of that year while staying at the home of Contessa Maggi, the original Mille Miglia's founder's wife. Another house visitor was Walter Hayes, VP of Portage of Europe. Notwithstanding issues over the past obtaining of AC Vehicles, Hayes saw the capability of the brand and the conversation brought about Portage taking an offer holding in September 1987.In 1988, having delivered nearly 5,000 vehicles in 20 years, a resuscitated economy and fruitful deals of restricted version Vantage, and 52 Volante Zagato roadsters at £86,000 each; Aston Martin at long last resigned the old V8 and presented the Virage range.

Despite the fact that Gauntlett was legally to remain as executive for a long time, his dashing advantages took the organization back into sports vehicle hustling in 1989 with restricted European achievement. Nonetheless, with motor rule changes for the 1990 season and the send off of the new Volante model, Portage gave the restricted stockpile of Cosworth motors to the Puma vehicles hustling group. Ford agreed to take full control of Aston Martin because the entry-level DB7 would need a lot of engineering work. In 1991, Gauntlett gave Hayes the chairmanship of Aston Martin. In 1992, Aston Martin introduced the Vantage, a high-performance version of the Virage, and the following year, it introduced the DB7.

After acquiring a stake in the business in 1987, Ford included Aston Martin in the Premier Automotive Group in 1993, invested in new manufacturing, and increased production. For the purpose of producing the DB7, Ford opened a brand-new plant in 1994 on Banbury Road in Bloxham. Aston Martin produced a record 700 vehicles in 1995. Before Ford, hand coachbuilding techniques like the English wheel were used to make cars. During the mid 1990s, the Extraordinary Undertakings Gathering, a mysterious unit with Works Administration at Newport Pagnell, made a variety of unique mentor constructed vehicles for the Brunei imperial family.] In 1998, the 2,000th DB7 was assembled, and in 2002, the 6,000th, surpassing creation of all of the past DB series models. The DB7 territory was redone by the expansion of all the more remarkable V12 Vantage models in 1999, and in 2001, Aston Martin presented the V12-engined lead model called the Vanquish which succeeded the maturing Virage (presently called the V8 Car).

At the North American Global Car expo in Detroit, Michigan in 2003, Aston Martin presented the V8 Vantage idea vehicle. Expected to have not many changes before its presentation in 2005, the Vantage carried back the exemplary V8 motor to permit Aston Martin to contend in a bigger market. The Gaydon factory, Aston Martin's first purpose-built facility, opened in 2003 as well. The office is arranged on a 55-section of land (22 ha) site of a previous RAF V Plane airbase, with a 8,000 m2 (86,000 sq ft) front structure for workplaces, meeting rooms and client gathering, and a 35,000 m2 (380,000 sq ft) creation building.

The DB9 coupé, which replaced the DB7, was also introduced in 2003. At the 2004 Detroit auto show, the DB9 Volante, a convertible version of the DB9, was unveiled.

In October 2004, Aston Martin set up the devoted 12,500 m2 (135,000 sq ft) Aston Martin Motor Plant (AMEP) inside the Portage Germany Niehl, Cologne plant. As with traditional Aston Martin engine production from Newport Pagnell, each unit was assembled by a single technician from a pool of 30, with V8 and V12 variants assembled in under 20 hours. The facility had the capacity to produce up to 5,000 engines annually by 100 specially trained personnel. By taking motor creation back to inside Aston Martin, the commitment was that Aston Martin would have the option to deliver little runs of better execution variations' motors. The entry-level V8 Vantage sports car was able to join the DB9 and DB9 Volante in production at the Gaydon factory in 2006 thanks to this increased engine capacity.

In December 2003, Aston Martin declared it would get back to engine hustling in 2005. Together with Prodrive, the design, development, and management of the DBR9 program were assigned to a new division known as Aston Martin Racing. The DBR9 contends in the GT class in sports vehicle races, including the undeniably popular 24 Hours of Le Monitors.

A Ford internal audit in 2006 prompted the company to consider selling off a portion of its Premier Automotive Group. Ford announced in August 2006 that it had engaged UBS AG to sell all or a portion of Aston Martin at auction after considering the possibility of selling Jaguar, Land Rover, or Volvo automobiles.

2007–2018: Private Limited Company On March 12, 2007, an American investment banker named John Sinders and two Kuwaiti businesses called Investment Dar and Adeem Investment joined forces with Prodrive chairman David Richards to purchase Aston Martin for £475 million (US$848 million). Prodrive had no monetary contribution in the arrangement. Ford maintained a £40 million (US$70 million) stake in Aston Martin.

To exhibit the V8 Vantage's solidness across risky landscape and advance the vehicle in China, the principal east-west going across of the Asian Thruway was embraced among June and August 2007. A couple of Britons traveled 12,089 km (7,512 miles) from Tokyo to Istanbul prior to joining the European motorway network for another 3,259 km (2,025 miles) to London. Aston Martin opened dealerships in Shanghai and Beijing within three months of the promotion's success.

On 19 July 2007, the Newport Pagnell plant carried out the remainder of almost 13,000 vehicles made there beginning around 1955, a Vanquish S. The Tickford Road office was changed over and turned into the home of the Aston Martin Works exemplary vehicle division which centers around legacy deals, administration, extras and reclamation tasks. The 55-acre (22 ha) Gaydon facility on the former RAF V Bomber airbase became the center of UK production.[81] In March 2008, Aston Martin announced a partnership with Magna Steyr to outsource the production of over 2,000 cars annually to Graz, Austria, stating reassuringly: " The proceeding with development and progress of Aston Martin depends on Gaydon as the point of convergence and heart of the business, with the plan and designing of all Aston Martin items proceeding to be done there."

The new pair in China and additional dealers in Europe brought the total to 120 in 28 countries. Aston Martin made the announcement on September 1, 2008, that the Lagonda brand would be revived and proposed a concept car that would be displayed in 2009 to coincide with the brand's 100th anniversary. In response to the economic downturn, Aston Martin announced in December 2008 that it would reduce the number of employees it employed from 1,850 to 1,250.

The initial four-entryway Rapide terrific travelers carried out of the Magna Steyr processing plant in Graz, Austria in 2010.The agreement maker furnishes devoted offices to guarantee consistence with the demanding principles of Aston Martin and different marques, including Mercedes-Benz. Then Chief of the organization, Ulrich Bez had openly guessed about rethinking Aston Martin's all's tasks except for marketing.[88] In September 2011, it was reported that development of the Rapide would be gotten back to Gaydon in the final part of 2012, reestablishing the organization's all's vehicle make there.

On December 6, 2012, the Italian private equity fund Investindustrial agreed to purchase a 37.5% stake in Aston Martin and invest £150 million as a capital increase. This was affirmed by Aston Martin in an official statement on 7 December 2012.David Richards left Aston Martin in 2013, getting back to focus on Prodrive.
Bez was said to be leaving his position as chief executive officer in April 2013 to take on a more ambassadorial role. On 2 September 2014, Aston Martin declared it

had designated the Nissan leader Andy Palmer as the new President with Bez holding a situation as non-chief director. As deals had been declining from 2015, Aston Martin looked for new clients (especially well off female purchasers) with presenting idea vehicles like the DBX SUV alongside track centered vehicles like the Vulcan. As per Palmer, the difficulties began when deals of the DB9 neglected to create adequate asset to foster cutting edge models which prompted a descending winding of declining deals and productivity.

Palmer framed that the organization intends to foster two new stages, add a hybrid, invigorate its supercar setup and influence its innovation coalition with Daimler as a feature of its six-year intend to make the 100-year-old English brand reliably beneficial. "We went bankrupt seven times in the first century," he stated. Changing that is the focus of the second century. "In anticipation of its up and coming age of sports vehicles, the organization contributed £20 million ($33.4 million) to grow its assembling plant in Gaydon. A new chassis and pilot build facility, an expanded storage area for parts and logistics, and brand-new offices are all part of the Gaydon plant's expansion. The plant will be expanded by Aston Martin by approximately 10,000 square meters (or 110,000 square feet).

In 2014, Aston Martin experienced a pre-charge deficiency of £72 million, practically triple of how much 2013 selling 3,500 vehicles during the year, well beneath the 7,300 vehicles sold in 2007 and 4,200 sold in 2013 separately. In addition to the £304 million of senior secured notes that were issued in 2011 at 9.25% interest, Aston Martin had to secure an additional £200 million investment from its shareholders to fund the development of new models. In March 2014, the company issued "payment in kind" notes for US$165 million. Aston Martin's pre-tax losses for 2016 were reported to have increased by 27% to £162.8 million, marking the sixth consecutive year of losses.

The company chose a 36-hectare (90-acre) St. Athan, South Wales, location for its new factory in 2016. Despite fierce competition from locations as far away as the Americas, Eastern Europe, the Middle East, and Europe, as well as two other sites in the UK, which are thought to be Bridgend and Birmingham, Aston's board unanimously selected the Welsh facility. The office included three existing 'super-overhangs' of MOD St Athan. Development work of changing over the shelters started in April 2017. After selling more than 5,000 cars in 2016, Aston Martin made a profit again in 2017. In comparison to its loss of £163 million in 2016, the company recorded a pre-tax profit of £87 million in 2017. 2017 also marked the ten-year anniversary of the Newport Pagnell facility's return to production.

2013–present: Partnership with Mercedes-Benz Group In December 2013, Aston Martin entered into a contract with Mercedes-Benz Group, then known as Daimler, to provide Mercedes-AMG engines for the next generation of Aston Martin automobiles. Aston Martin was also going to receive electrical systems from Mercedes-AMG. Aston Martin wanted to use this technical partnership to help them launch a new line of cars with new engines and technology. Mercedes will receive a non-voting seat on Aston Martin's board and up to 5% equity in the company. The principal model to brandish the Mercedes-Benz innovation was the DB11, declared at the 86th Geneva Engine Show in Walk 2016. It included Mercedes-Benz hardware for the amusement, route and different frameworks. Additionally, it was the first model to utilize Mercedes-AMG V8 motors. In October 2020, Mercedes affirmed it will build its holding "in stages" from 5% to 20%.In return, Aston Martin will approach Mercedes-Benz half and half and electric drivetrain advances for its future models.

2018–present: Recorded on the London Stock Trade

In the wake of "finishing a circle back for the once perpetually misfortune making organization that could now be esteemed at as much as 5 billion pounds ($6.4 billion)," and presently revealing an entire year pre-charge benefit of £87 million (contrasted and a £163 million misfortune in 2016) Aston Martin in August 2018 declared plans to drift the organization at the London Stock Trade as Aston Martin Lagonda Worldwide Possessions plc. The organization was the subject of a first sale of stock on the London Stock Trade on 3 October 2018.In that very year, Aston Martin opened another vehicle elements test and improvement focus at Silverstone's Stowe Circuit close by another HQ in London. The company opened its new 36 ha (90 acres) factory in St. Athan in June 2019 to produce its first SUV, the DBX. On December 6, 2019, the factory was finally finished and opened for business. The factory will employ approximately 600 people when full production begins in the second quarter of 2020, and 750 when peak production is reached.

On January 31, 2020, it was announced that Yew Tree Overseas Limited, led by Canadian billionaire and investor Lawrence Stroll, would pay £182 million for a 16.7% stake in the company. The re-organizing incorporates a £318 million money implantation through another privileges issue, producing a sum of £500 million for the organization. Walk will likewise be named as director, supplanting Penny Hughes. Swiss drug tycoon Ernesto Bertarelli and Mercedes-AMG Petronas F1 group head and President Toto Wolff have additionally joined the consortium, obtaining 3.4% and 4.8% stakes, separately. Stroll increased his stake in the business by 25% in March 2020.

Aston Martin announced Andy Palmer's resignation as CEO on May 26, 2020. Tobias Moers of Mercedes-AMG will succeed him beginning 1 August, with Keith Stanton as interval head working official. As a result of the COVID-19 pandemic lockdown, the company announced in June 2020 that it would lay off 500 employees due to weak sales. In Walk 2021, leader administrator Lawrence Walk expressed that the organization anticipates building electric vehicles by 2025.In May 2022, Aston Martin named 76-year old Amedeo Felisa as the new CEO, supplanting Tobias Moers. Roberto Fedeli was additionally reported as the new boss specialized official.

In November 2020, a correspondences organization called Clarendon Interchanges distributed a report looking at the ecological effect of different powertrain choices for vehicles. After the report got inclusion from The Sunday Times and different distributions, it arose that the organization had been set up in February that year and was enlisted under the name of Rebecca Stephens - the spouse of James Stephens, who is the public authority undertakings head of Aston Martin Lagonda. According to the report, which cited a study by Polestar, electric vehicles would need to be driven for 48,000 miles (77,000 kilometers) before they would emit less CO2 than a gasoline-powered vehicle. Electric vehicle researcher Auke Hoekstra disputed this assertion, arguing that the report underestimated the emissions from vehicles with combustion engines and did not take into account the emissions from the production of gasoline. He claims that a typical electric vehicle would need to travel between 16,000 and 18,000 miles (25,700 and 30,000 kilometers) to offset manufacturing emissions. The report also involved a number of other businesses, including Bosch.

Saudi Arabia's Public Investment Fund (PIF) will acquire two board seats in the business in July 2022 through a £78 million equity placement and a £575 million separate rights issue. After the rights issue, the Saudi fund will own 16.7% of Aston Martin, followed by 18.3% from Stroll's Yew Tree consortium and 9.7% from the Mercedes-Benz Group. Chinese automaker Geely will buy a 7.6% stake in the company in September 2022. Stroll and the Yew Tree consortium increased their stake in the business from 28.29% in December 2022 to 17% in May 2023, making Geely the third-largest shareholder after the Saudi Arabia Public Investment Fund and the Yew Tree consortium.

In June 2023, Aston Martin consented to an arrangement with Clear Engines in the wake of choosing it to assist with providing electric engines, powertrains, and battery frameworks for its impending scope of completely electric vehicles. Consequently, Aston Martin will make cash installments and issue a 3.7 percent stake in its organization to Clear, worth $232 million altogether. In October 2023, Aston Martin declared that it would contend in the FIA World Perseverance Title and IMSA SportsCar Title in 2025.

Remarkable occasions

In August 2017, a 1956 Aston Martin DBR1/1 sold at a Sotheby's closeout at the Stone Ocean side, California Concours d'Elegance for US$22,550,000, which made it the most costly English vehicle at any point sold at a sale, as per Sotheby's. Stirling Moss and Carroll Shelby had previously been behind the wheel of the vehicle. A 1962 Aston Martin DB4 GT Zagato sold for US$14,300,000 at an auction in New York in 2015, and a 1963 Aston Martin DP215 sold for US$21,455,000 in August 2018.

Chapter 2:Aston Martin models

It was only after 1915 that Bamford and Martin would make the principal vehicle completely created by the organization, however creation was immediately stopped by WWI. After the conflict, Bamford and Martin took up creation at an area in Kensington, London, where it would deliver roughly 55 vehicles before the organization failed in 1924. In 1925, the company was purchased by Bill Renwick, Augustus Bertelli, and a group of investors, so there was no other option but to sell it. Aston Martin Motors was given a new name by the new owners in the same year. By 1932, the organization once more confronted monetary challenges and was offered to an individual named Spear Prideaux Brune, who then gave it to Sir Arthur Sutherland.

The very first automobile, codenamed "Coal Scuttle," was registered in 1915; a second vehicle was not completed until 1920, following the conclusion of the Great War. AM initially produced competitive 1500cc light car racers, but despite their track success, production and sales were sluggish, and Bamford & Martin entered receivership in 1925. Only 61 automobiles were produced at this time, of which 25 are known to have survived to the present day.

Pre-war cars

1921–1925 Aston Martin Standard Sports

1927–1932 Aston Martin First Series

1929–1932 Aston Martin International

1932–1932 Aston Martin International Le Mans

1932–1934 Aston Martin Le Mans

1933–1934 Aston Martin 12/50 Standard

1934–1936 Aston Martin Mk II

1934–1936 Aston Martin Ulster

1936–1940 Aston Martin 2-litre Speed Models (23 built) The last 8 were fitted with C-type bodywork

1937–1939 Aston Martin 15/98

Post-war cars

1948–1950 Aston Martin 2-Litre Sports (DB1)

1950–1953 Aston Martin DB2

1953–1957 Aston Martin DB2/4

1957–1959 Aston Martin DB Mark III

1958–1963 Aston Martin DB4

1961–1963 Aston Martin DB4 GT Zagato

1963–1965 Aston Martin DB5

1965–1966 Aston Martin Short Chassis Volante

1965–1969 Aston Martin DB6

1967–1972 Aston Martin DBS

1969–1989 Aston Martin V8

1977–1989 Aston Martin V8 Vantage

1986–1990 Aston Martin V8 Zagato

1989–1996 Aston Martin Virage/Virage Volante

1989–2000 Aston Martin Virage

1993–2000 Aston Martin Vantage

1996–2000 Aston Martin V8 Coupe/V8 Volante

1993–2003 Aston Martin DB7/DB7 Vantage

2001–2007 Aston Martin V12 Vanquish/Vanquish S

2002–2003 Aston Martin DB7 Zagato

2002–2004 Aston Martin DB AR1

2004–2016 Aston Martin DB9

2005–2018 Aston Martin V8 and V12 Vantage

2007–2012 Aston Martin DBS V12

2009–2012 Aston Martin One-77

2010–2020 Aston Martin Rapide/Rapide S

2011–2012 Aston Martin Virage/Virage Volante

2011–2013 Aston Martin Cygnet, based on the Toyota iQ

2012–2013 Aston Martin V12 Zagato

2012–2018 Aston Martin Vanquish/Vanquish Volante

2015–2016 Aston Martin Vulcan

2016–2023 Aston Martin DB11

2018–present Aston Martin Vantage

2018–present Aston Martin DBS Superleggera

2020–present Aston Martin DBX

2023–present Aston Martin DB12

Other

1944 Aston Martin Atom (concept)

1961–1964 Lagonda Rapide

1976–1989 Aston Martin Lagonda

1980 Aston Martin Bulldog (concept)

1993 Lagonda Vignale (concept)

2001 Aston Martin Twenty Twenty (Italdesign concept)

2007 Aston Martin V12 Vantage RS (concept)

2007–2008 Aston Martin V8 Vantage N400

2009 Aston Martin Lagonda SUV (concept)

2010 Aston Martin V12 Vantage Carbon Black Edition

2010 Aston Martin DBS Carbon Black Edition

2013 Aston Martin Rapide Bertone Jet 2+2 (concept)

2013 Aston Martin CC100 Speedster (concept)

2015 Aston Martin DB10 (concept)

2015–2016 Lagonda Taraf

2019 Aston Martin Vanquish Vision (concept)

2019 Aston Martin DBS GT Zagato

2020 Aston Martin V12 Speedster

2021 Aston Martin Victor

2022 Aston Martin DBR22

2023 Aston Martin Valour

Current models

Aston Martin DB12

Aston Martin DBS Superleggera

Aston Martin DBX

Aston Martin Vantage

Aston Martin Valkyrie

Upcoming models

Aston Martin Valhalla

Aston Martin Lagonda All-Terrain

WHY IS ASTON MARTIN CALLED DB?

In 1947, David Brown Limited, a tractor manufacturer, bought the company while collaborating with his other automotive business, Lagonda. Lagonda is a significant piece of the riddle as David was dealing with another V6 motor that turned over the '2 liter games' (named the DB). Because of this new progression Aston Martin models started to likewise brandish David Earthy colored's initials:

1950-1953: DB2 (410 built)

1953-1957: DB2/4 (761 built)

1957-1959: DB MkIII (552 built)

1958-1963: DB4 (1185 built)

1960-1962: DB4 Zagato (19 built)

1963-1965: DB5 (1059 built)

1965-1950: DB6 (1788 built)

1967-1972: DBS (1193 built)

The DB models become notable subsequent to showing up in various James Bond films over the course of the years to follow. In 1972 David Brown sold the organization.

ASTON MARTIN VANTAGE, V8, AND VOLANTE

While the DB designation initially featured a brand-new V6 engine, additional Aston Martin models were in the works. This soar another period of exceptional vehicles:

1969: V8 version of the DBS

1972: unveiling of AMV8

1977: new Aston Martin Vantage – hailed as the first British supercar

1980: a concept car called Bulldog

1987: Ford Motors took a substantial holding in the company

1988: V8 Virage made its debut

ASTON MARTIN AND FORD MOTORS When Ford Motors took over, production of Aston Martin automobiles increased while maintaining the marque's distinctive quality reputation. We also witnessed the DB7 redesign of database models. During this time new models were delivered which included:

1988-1996: V8 Virage (598 built)

1993-2000: V8 Vantage (288 built)

1994-1999: DB7 (2,461 built)

1996-2000: V8 Coupe (101 built)

1997-2000: V8 Volante (64 built)

1999-2003: DB7 Vantage (4,431 built)

1999-2003: DB7 Vantage Voltage (2,046 built)

2001-2007: V12 Vanquish (2,578 built)

2003: DB7 Zagato (200 built)

Aston Martin and James Bond maintained a strong connection during this Ford era, as numerous models continued to be featured in these films. Mercedes-AMG and Aston Martin formed a partnership in 2013.

Chapter 3: Why an Aston Martin is so desirable

For more than a century, the luxury car brand Aston Martin has been associated with style, elegance, and high performance. A brand has caught the hearts of vehicle fans and extravagance enthusiasts the same with its smooth plan, strong motor, and rich legacy. Aston Martin remains one of the most sought-after luxury car brands despite the numerous brands available on the market. This book delves into the history, design philosophy, and engineering prowess of Aston Martin to find out why the brand is so coveted.

Aston Martins are renowned for their timeless and elegant design, constructed from high-quality materials like carbon fiber and leather. The brand stands out from other manufacturers of luxury automobiles thanks to its distinctive appearance. Aston Martin vehicles are highly sought after by car enthusiasts due to their sporty and sophisticated lines and curves.

RICH HISTORY Aston Martin has been around for more than a century. Since the 1960s, the brand has been associated with James Bond films, establishing its status as a symbol of luxury and sophistication. Aston Martin's connection to motorsports is an important part of its history. The brand has been involved in racing for a long time, and its automobiles have participated in some of the world's most prestigious races, such as the Le Mans 24 Hours and the Formula One World Championship.

UNBEATABLE PERFORMANCE Due to the brand's emphasis on superior engineering, its automobiles exhibit impressive acceleration, speed, and handling. The Aston Martin Vantage GTE, which competes in the FIA World Endurance Championship, is a good example of this. With advanced aerodynamics, suspension systems, and other technologies, these automobiles are made to perform at their absolute best on the racetrack.

One-Of-A-Kind Exclusion Aston Martin makes only a few cars a year, making them more exclusive and more difficult to obtain than other luxury cars. The brand has a long history of creating very good quality games vehicles extremely popular among vehicle lovers, meaning it can stand to be specific about who purchases its vehicles. Moreover, Aston Martin offers a scope of customization choices that permit clients to make totally tailor made vehicles, adding to the restrictiveness, as every vehicle is extraordinary.

The v8 vantage, for instance, is the cheapest Aston Martin. The price is $123,695. That money gets you a shape that is heartbreakingly beautiful and an interior that is quietly elegant, but you might lose a few stoplight drags to a guy driving a new Porsche 911. You could have bought a brand-new 911 for $123,695. The details of the vantage differ from one another: a crystal starter button, a leather interior that looks like it was made by magic elves, and a level of finish that looks like it was made by humans, not robots. an engine that sounds like a combination of a machine gun battery and a crackling fire. instruments that make expensive watches appear like trash from a flea market.

All things considered, with extravagance vehicles, nuance is a specialty. In 2014, Aston only moved 4,000 sports cars and sedans around the world. To put things in perspective, Ferrari sells 7,000 of its hyper-focused exotics annually and has voluntarily limited sales for a long time. In order to safeguard Ferrari's long-term brand health, the company has successfully managed its expansion over many years. This is in contrast to Porsche, where SUVs now account for more than half of the company's global volume of 189,000 vehicles. The number of sports cars manufactured by the German company has never been higher. A long time back, before Porsche broadened its contributions and reexamined its creation techniques, its deals were scarcely above where Aston Martin is currently.

Chapter 4:Facts about Aston Martin

In 1913, Aston Martin was established.

With a rich history crossing north of hundred years, Aston Martin has become inseparable from extravagance, tastefulness, and immortal English craftsmanship.

The mythical creature's wings are featured in the Aston Martin logo.

A pair of wings, which stand for speed and independence, are depicted in the enduring Aston Martin logo. It is propelled by the winged logo of Bentley, which the organization utilized in its initial days.

An Aston Martin is James Bond's favorite automobile.

Since the film "Goldfinger" in 1964, Aston Martin has been firmly connected with the world's most popular fictitious government agent, James Bond. The English spy has driven different models of Aston Martin all through the film series, adding to the brand's social importance and prominence.

Aston Martin has a maximum velocity of north of 200 mph.

Aston Martin automobiles are renowned for their exceptional performance and can travel at incredible speeds. Some models, like the Aston Martin DBS Superleggera, can go faster than 200 mph, making driving them thrilling.

One of the most costly Aston Martin vehicles at any point sold was a DBR1.

With a staggering $22.5 million, a 1956 Aston Martin DBR1 became the most expensive British car ever sold at auction in 2017. The DBR1 is an exemplary dashing vehicle, praised for its excellence, extraordinariness, and motorsport accomplishments.

Aston Martin's leader model is the DB11.

The DB11 addresses the embodiment of Aston Martin's fantastic visiting legacy. With its smooth plan, strong motor, and sumptuous inside, the DB11 encapsulates the ideal mix of execution and solace.

Red Bull Racing and Aston Martin have formed a partnership.

Aston Martin has collaborated with the Recipe 1 group Red Bull Dashing to foster trend setting innovations and elite execution vehicles. They teamed up on the Aston Martin Valkyrie, a hypercar intended to push the limits of auto designing.

Cars made by Aston Martin are made by hand.

Each Aston Martin car is hand-built with care by skilled craftsmen using both old-fashioned and new technology. Every vehicle receives the highest possible level of precision and quality as a result of this attention to detail.

The most well-known automobile in the world is the Aston Martin DB5.

Because of its appearance in the James Bond films, especially in "Goldfinger" and "Skyfall," the Aston Martin DB5 has gotten its status as a notorious and amazing vehicle, respected via vehicle fans and moviegoers the same.

The Aston Martin Valkyrie can be driven on the street.

The Valkyrie from Aston Martin is more than just a race car; being driven on the road is likewise planned. With its streamlined shape, lightweight development, and strong motor, it conveys an unequaled driving encounter.

Aston Martin has its own customization administration called "Q by Aston Martin."

A bespoke customization service known as "Q by Aston Martin" is available to customers who want a truly individualized Aston Martin. This help permits clients to tailor each part of their vehicle, from the outside paint tone to the inside trim and materials.

From 1958 to 1963, Aston Martin produced the iconic DB4.

The Aston Martin DB4 is generally viewed as an exemplary plan, highlighting a smooth and immortal outline. It represented a significant technological and performance advancement for the business.

Aston Martin has a long history in motorsports.

Aston Martin has serious areas of strength for an in motorsport, contending in different dashing series all over the planet. The company has won numerous competitions, including the World Endurance Championship and the 24 Hours of Le Mans overall championship.

Aston Martin teamed up with Italian plan house Zagato to make unique release models.

Aston Martin and Zagato have collaborated to create one-of-a-kind and highly sought-after automobiles. The combined expertise and aesthetic flair of both brands are on display in these limited-edition models.

Aston Martin's popular model, the Vantage, has been underway starting around 2005.

The Aston Martin Vantage is a sports car that is well-known for its striking design and dynamic performance. Because it strikes the perfect balance between athleticism and luxury, it has become one of the brand's most popular models.

Collectors have a high demand for Aston Martin automobiles.

Aston Martin automobiles are highly sought after by car enthusiasts and collectors worldwide due to their timeless elegance, exceptional performance, and limited production numbers. Claiming an Aston Martin isn't just an image of extravagance yet additionally an explanation of refined taste.

The Aston Martin Rapide is a four-entryway sports vehicle.

The Aston Martin Rapide provides a one-of-a-kind driving experience by combining the practicality of a four-door sedan with the exhilarating performance of a sports car. It stands out in the competitive luxury car market thanks to its sleek design and luxurious interior.